Journey of the Dream Team

Story & Photographs by

Sarah E. Grote

DEDICATION

I have authored this book in honor of Spencer and Oliver, and the incredible organization that has changed my life, the Connecticut Draft Horse Rescue, founded by Dr. Stacey Golub.

Extra special thanks go to my sister, Jen, and the love of my life, Gary. Their love and support have helped to make my own creative journey a reality, and many dreams come true. I am eternally grateful for my parents who instilled in me a love of nature, advocacy, a strong work ethic, and aesthetic beauty, as well as a commitment to excellence, kindness, and integrity.

Kudos to the incredible CDHR volunteers who have hearts of gold, incredible strength, and dedication to our horses and organization.

This book, "Journey of the Dream Team", was made possible with the support of the Department of Economic and Community Development, Office of the Arts, which also receives support from the federal ARPA.

First Printing, 2022

"I wish our old friend didn't leave us," said Oliver to Spencer.

"Me, too. Maybe he will send angels to help," said Spencer.

As weeks went by, the hay got moldier, the water trough turned blacker, and was the demise of a bird.

With minimal nutrition and access to only unclean water, their health was impacted dramatically.

"We sure need those angels," said Oliver.
"They will come, friend," said Spencer.

And one day, the two with hearts of gold saw them and were determined to save them.

"They'll find a way," said Spencer.

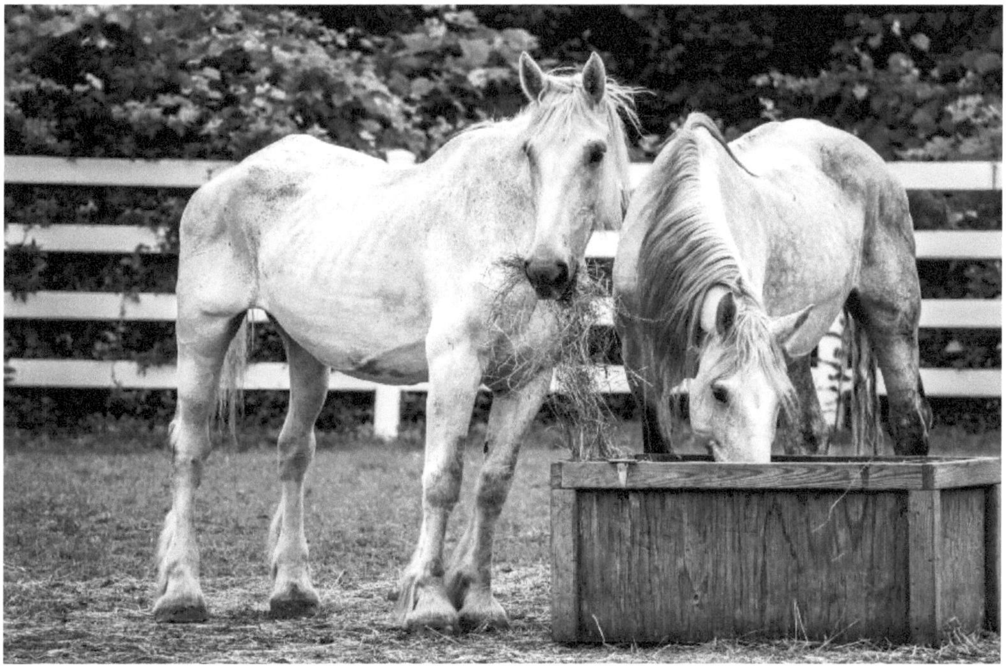

"We haven't been on a trailer in a long time. What if we are too skinny and weak?" asked Oliver.

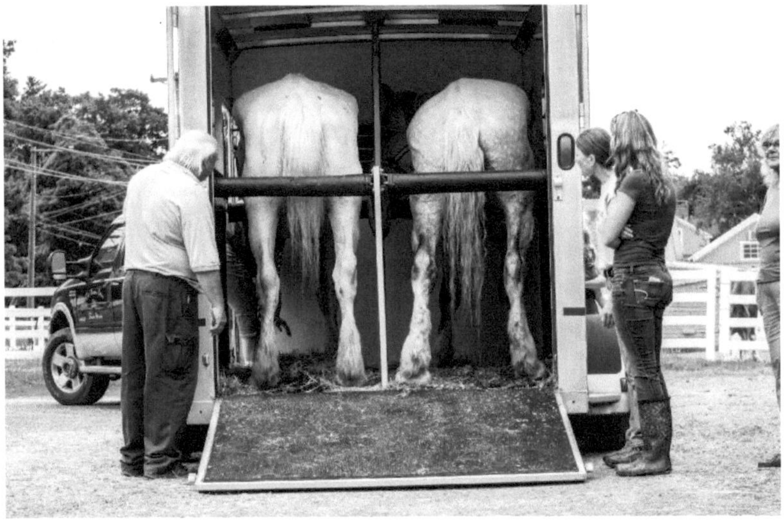

"Have faith, we must go to a nicer place. And make sure they bring Lila" Spencer said with a deep cough and sad eyes.

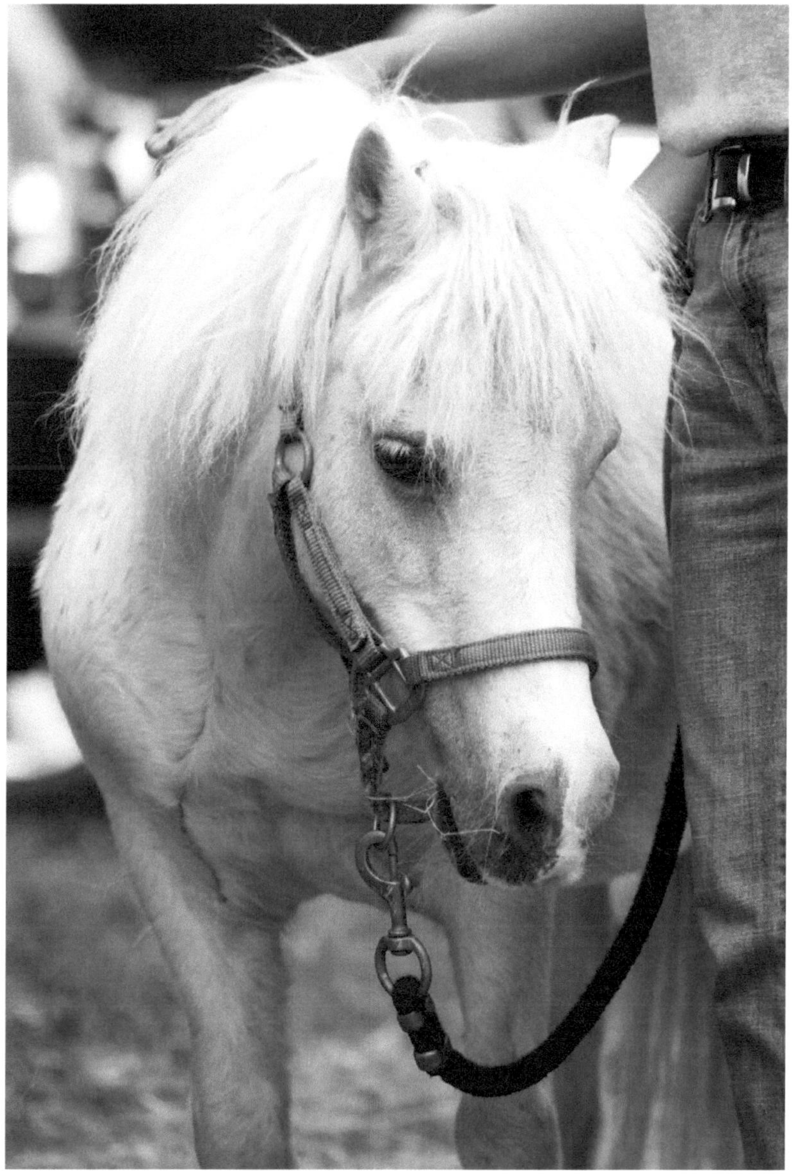

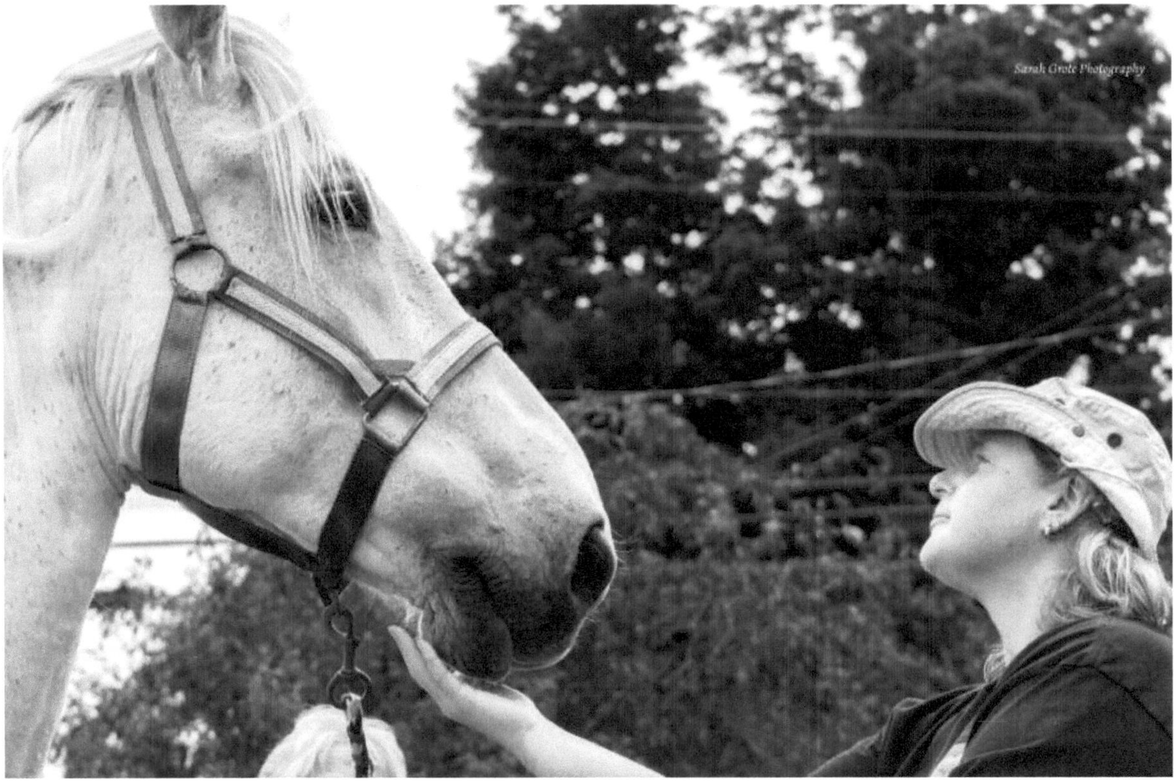

"Look at all of these people with tears in their eyes and open arms. We are going to be okay" said Spencer to Oliver.

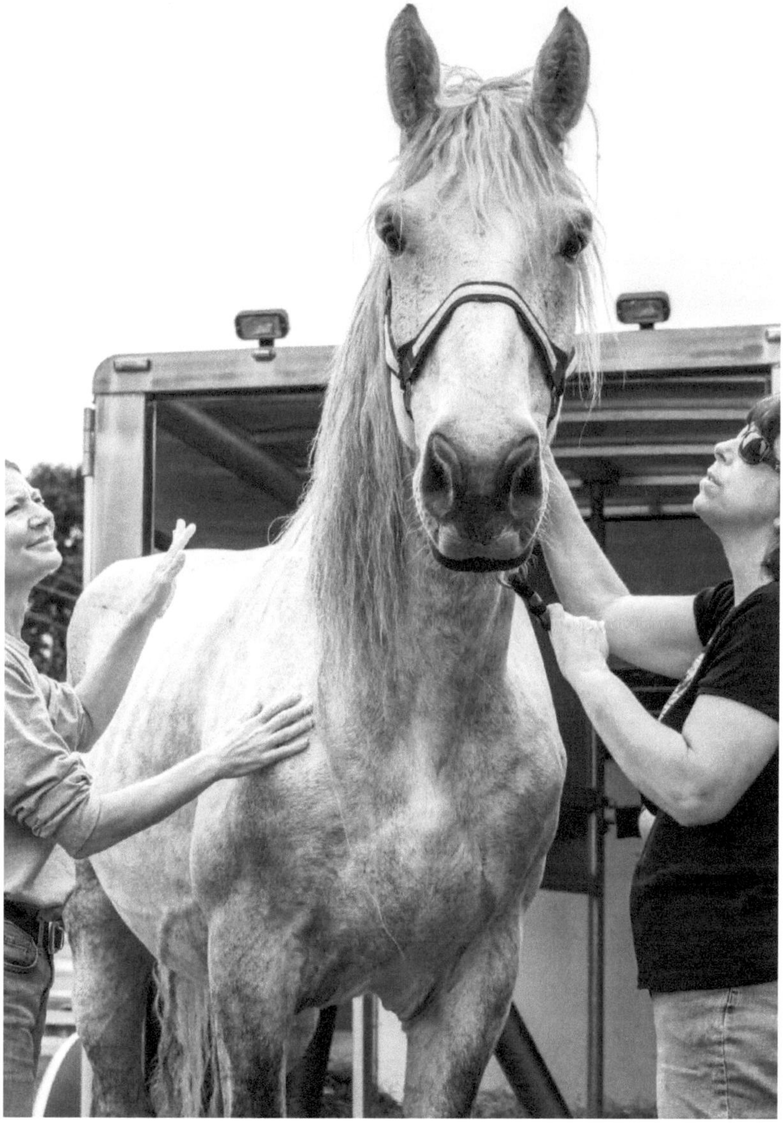

"Oh my gosh, a kind old man is bringing us hay. It looks so lush and tasty," said Oliver.

"Maybe our old man sent him from Heaven. We must be patient and respectful in honor of him," said Spencer.

As weeks went by, they enjoyed an abundance of glorious hay, and eventually got the grain and supplements, all served with an abundance of love by so many new friends.

"I feel stronger," they both said to each other in unison.

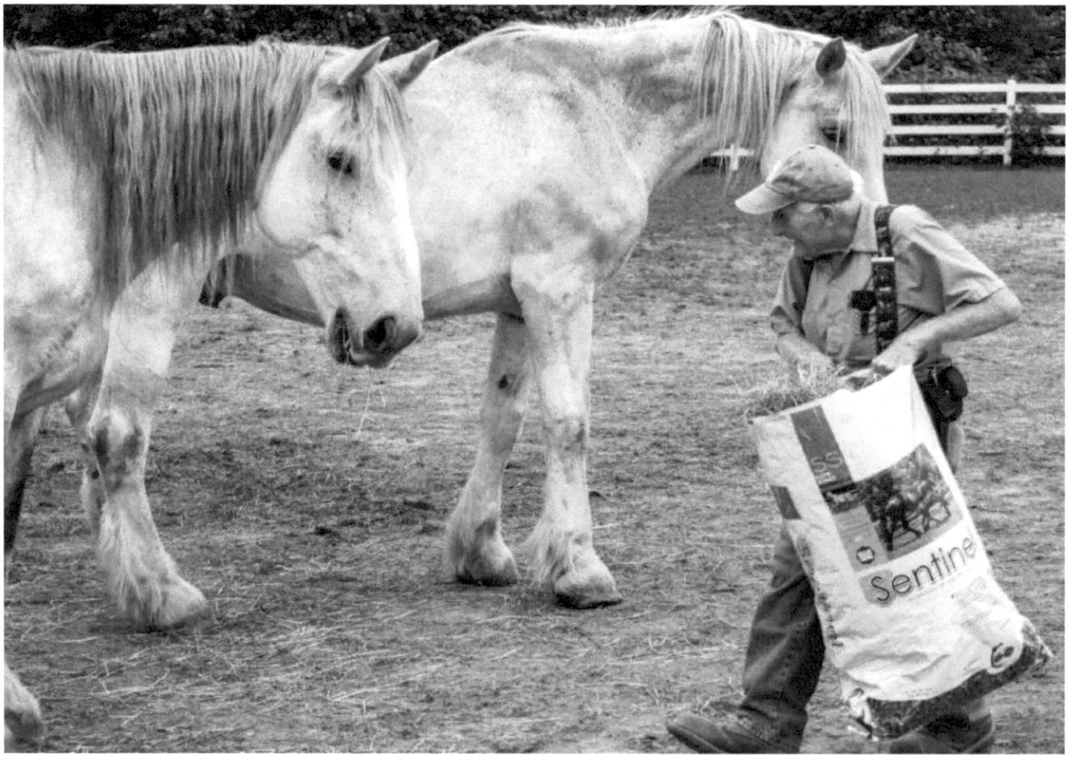

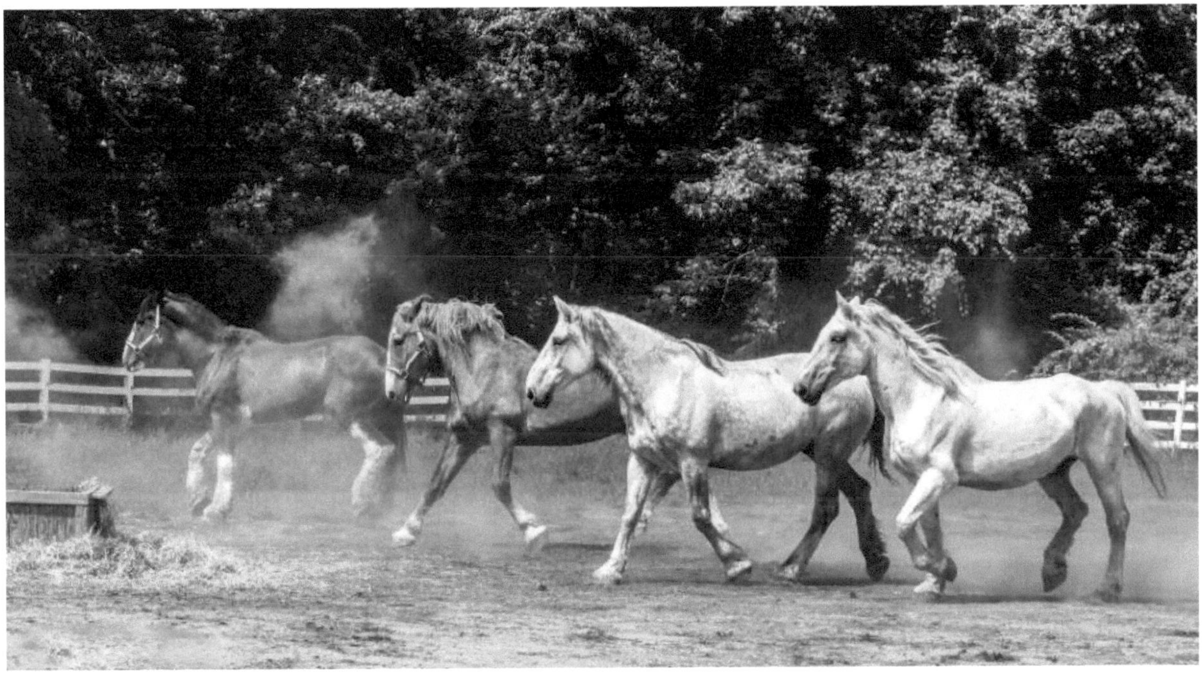

They also became friends with other horses on the farm. Some became paddock mates, while others trumpeted and whinnied from afar.

Duke the Clydesdale and Vern the Belgian became paddock mates with Spencer and Oliver that summer and were truly poetry in motion together.

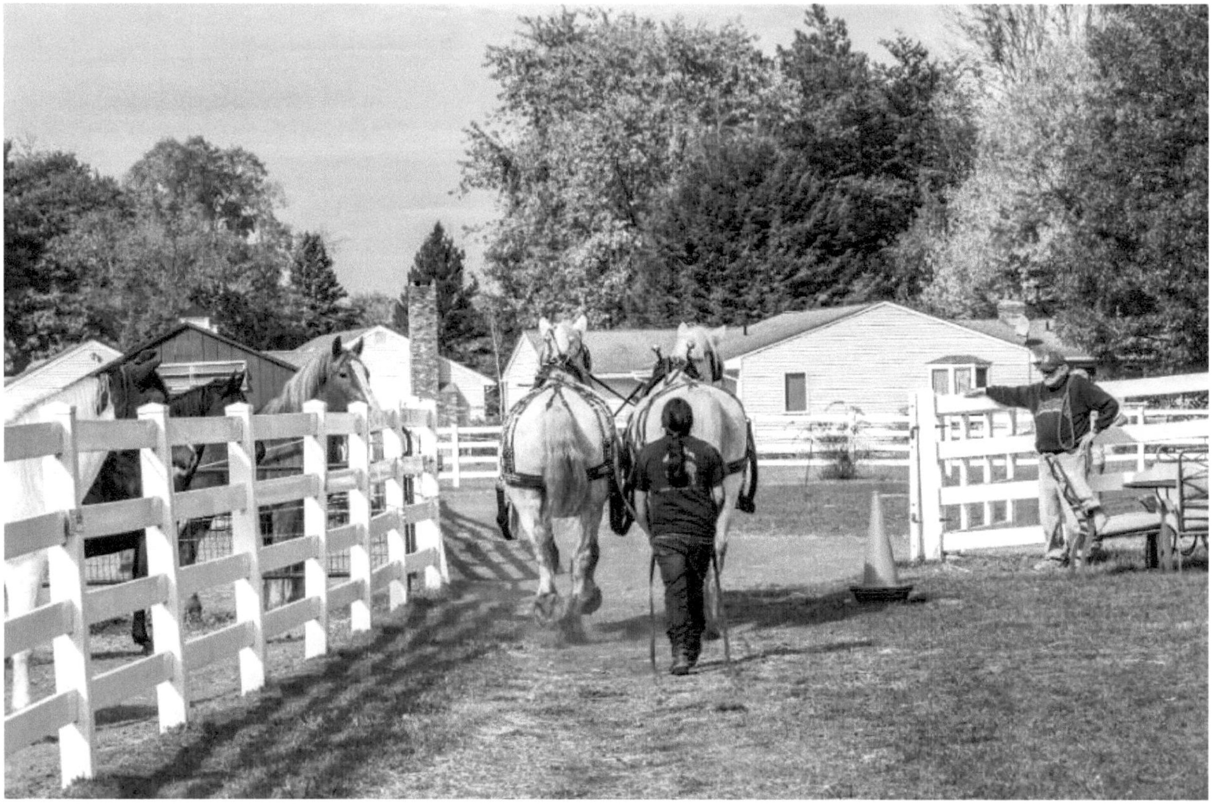

Then one day in early fall, they heard a familiar sound: the clanging of tack.

Instantly, their eyes brightened, remembering the parades and wagon rides when they were the dream team.

"Hitch them up to the wagon!" the great golden-haired doctor exclaimed with a smile that went from ear to ear.

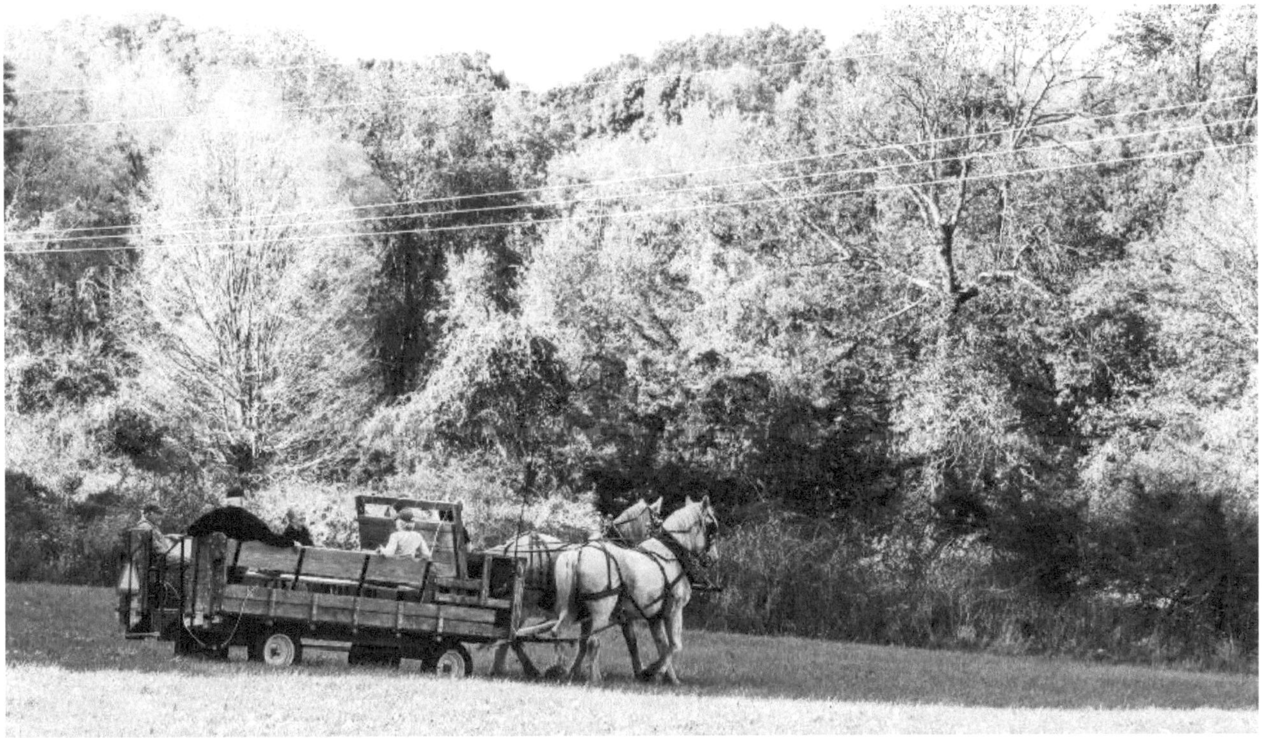

And there they were in perfect unison, giving their human friends a ride with pride, effortlessly transporting them along the gentle waves of grass in the hayfield, while the tree line beamed with a rainbow of fall colors.

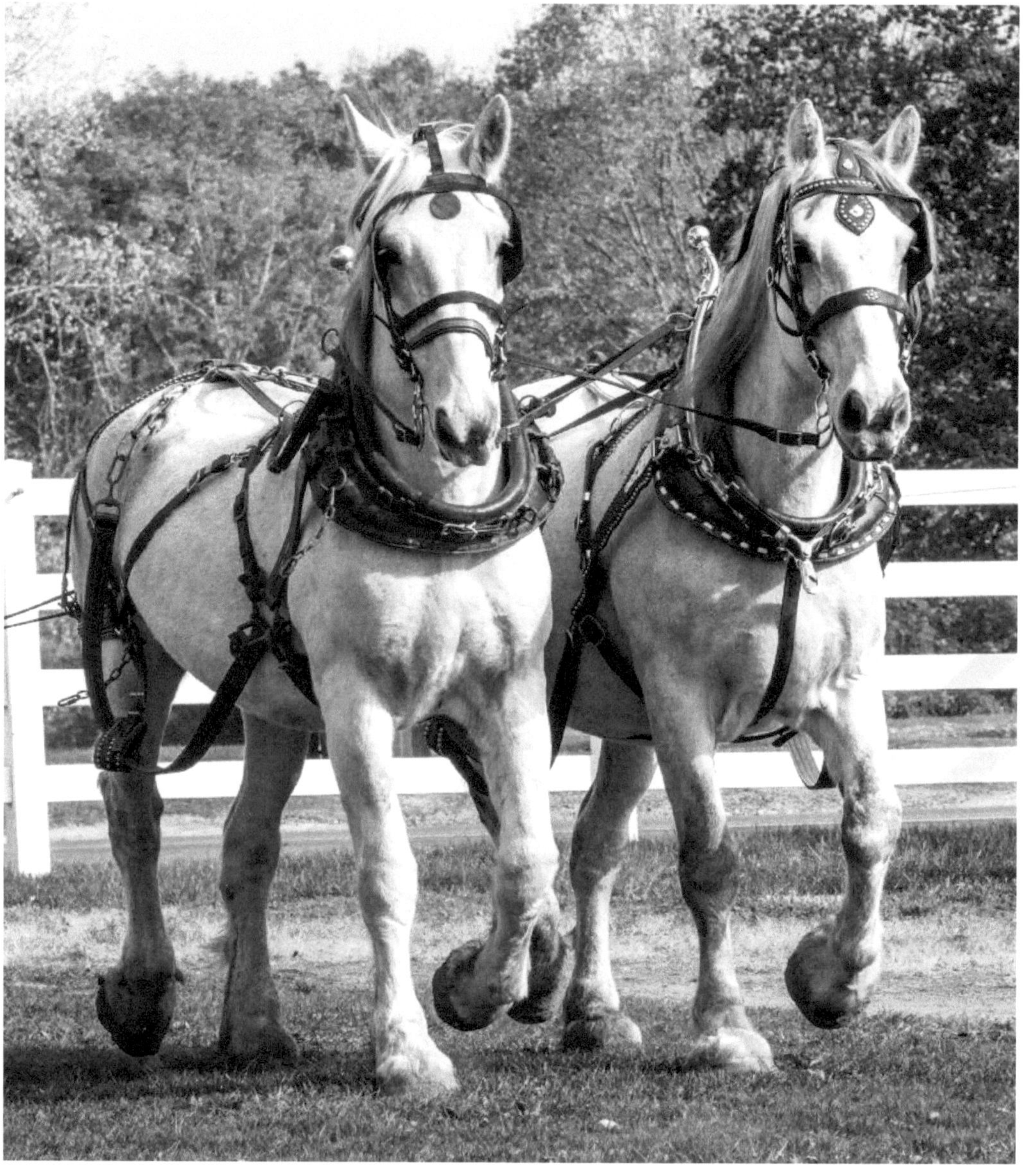

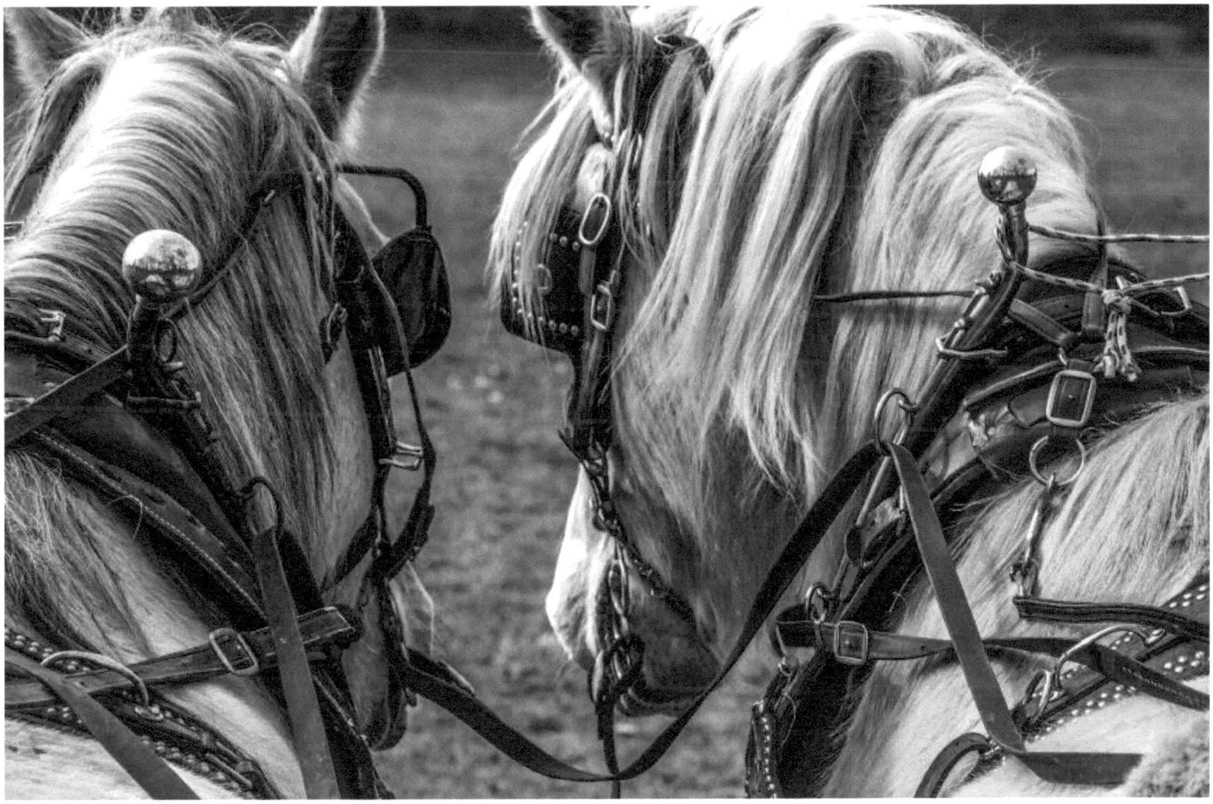

"We are so proud of you both. We proclaim you the ambassadors" the kind doctor told them. In typical stoic draft horse fashion, their reaction was subtle externally. But inside their huge hearts, they felt they now had crowns and were knighted.

They even gave their human friends rides on their backs. "They even give us kisses and a sweet treat. They must really like us," said Oliver.

One day, they noticed their human friends were crying, but they were not happy tears this time. Their paddock mate had told them he was not feeling good, and one sunny day he was not moving, while the people held him, braided his mane while decorating it with clovers.

"We must make them happy again," Spencer said to Oliver. "Yes, we must" Oliver replied.

"Let's drive them to the fair!" they both exclaimed exuberantly.

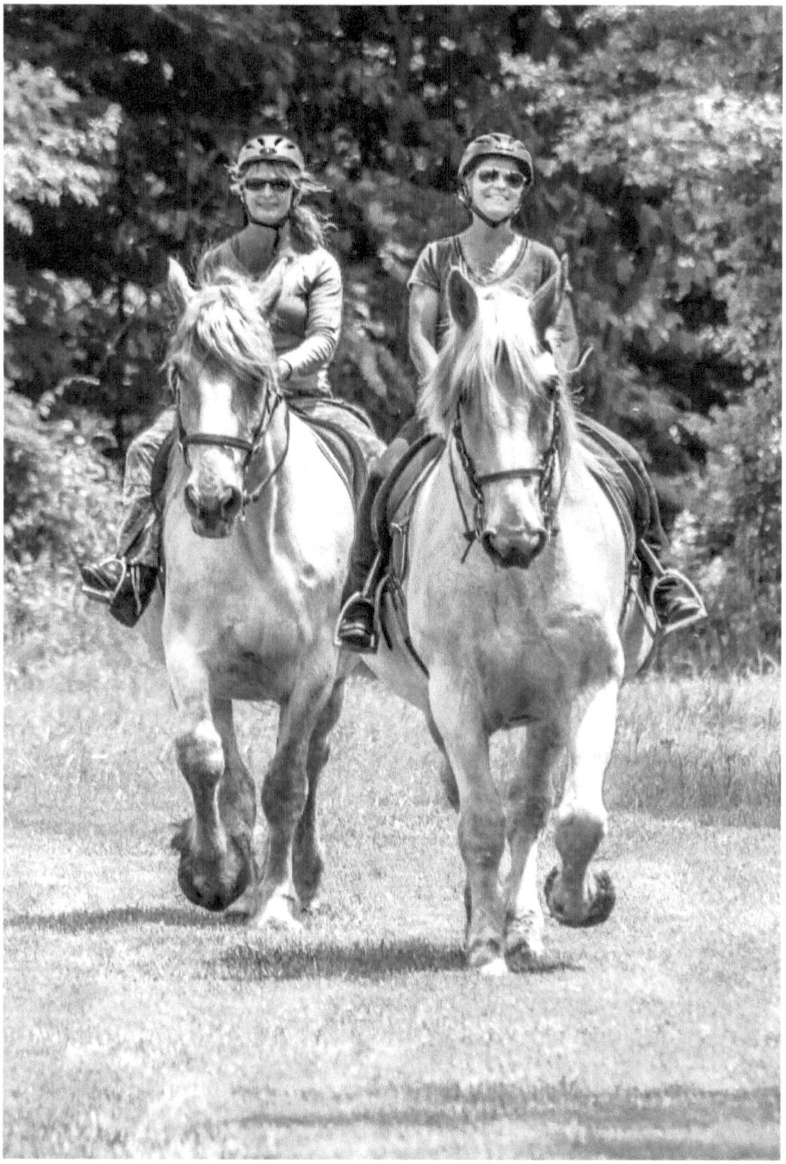

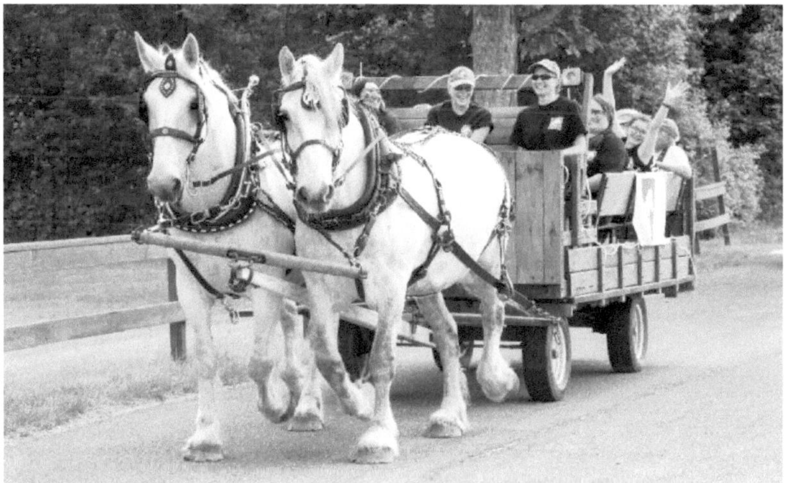

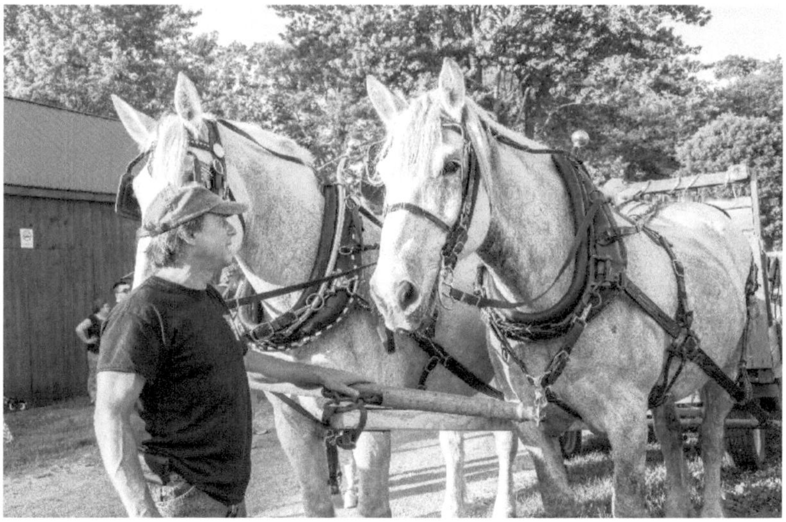

"I will take them snow tubing" trumpeted Spencer.

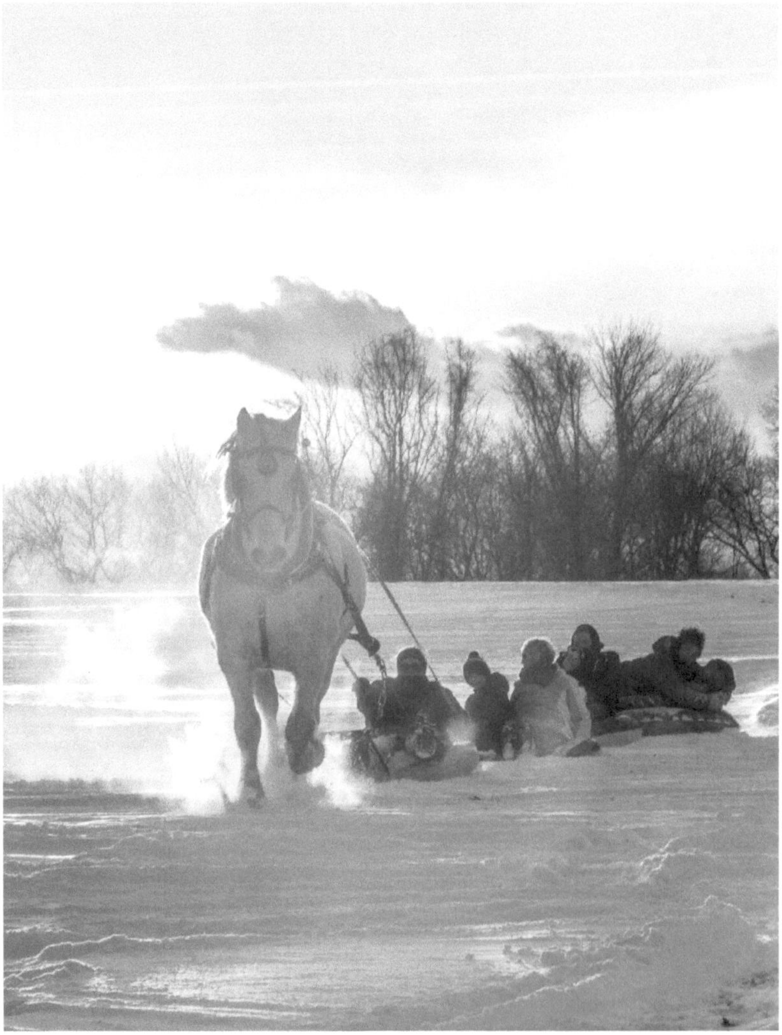

"Let's give our angels an anniversary ride," said Oliver with a smile.

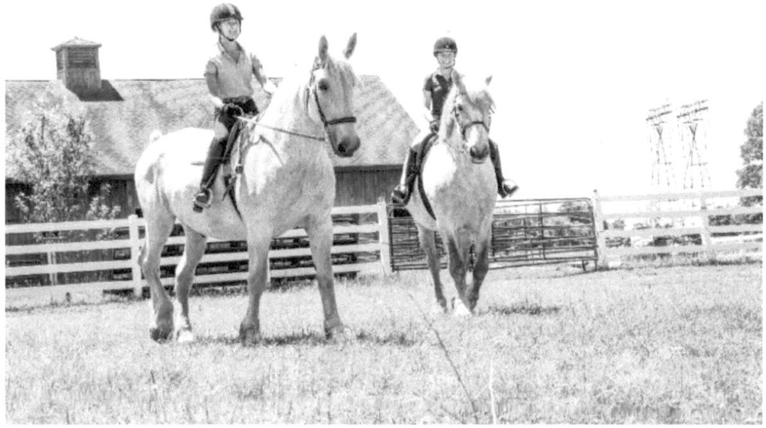

"Let's ride tandem to really make them happy!" said Oliver. They saw the spectators exclaim "Wow!". They of course got hugs, smiles, and kisses after.

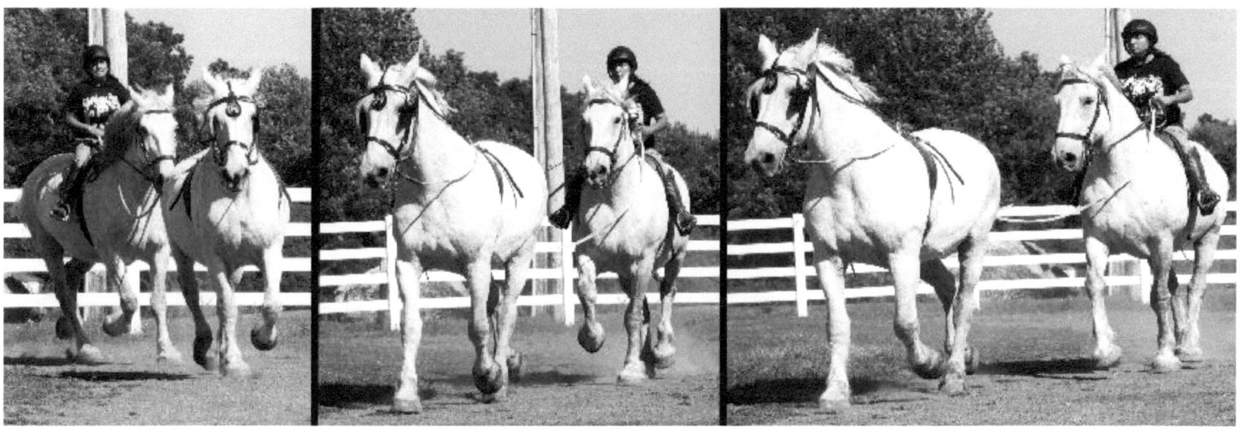

Then one day, the news came that they were going to a new farm.

"What about all of our friends, horses and humans alike?" a worried Vern asked.

"It's okay, Vern. Don't be nervous. We are all here together" Spencer gently told him.

"Phew, I sure love all of my friends" replied Oliver.

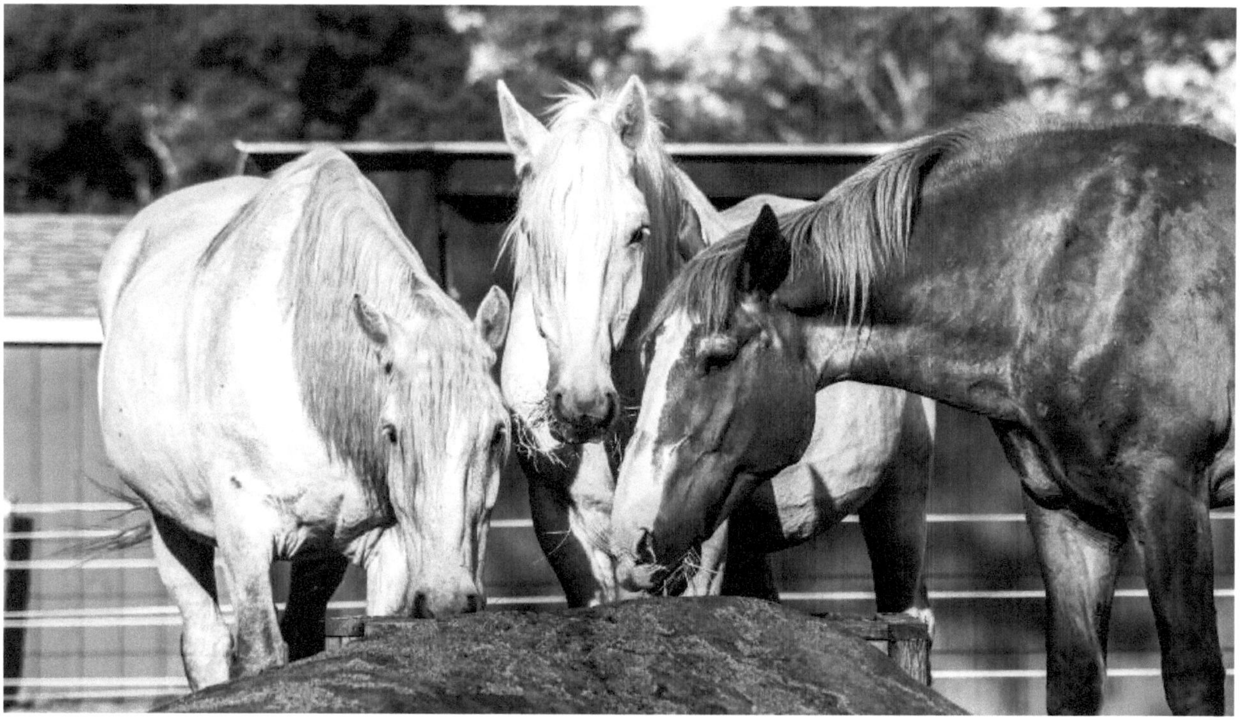

"Wow, look at that beautiful grand tree" gasped Oliver.

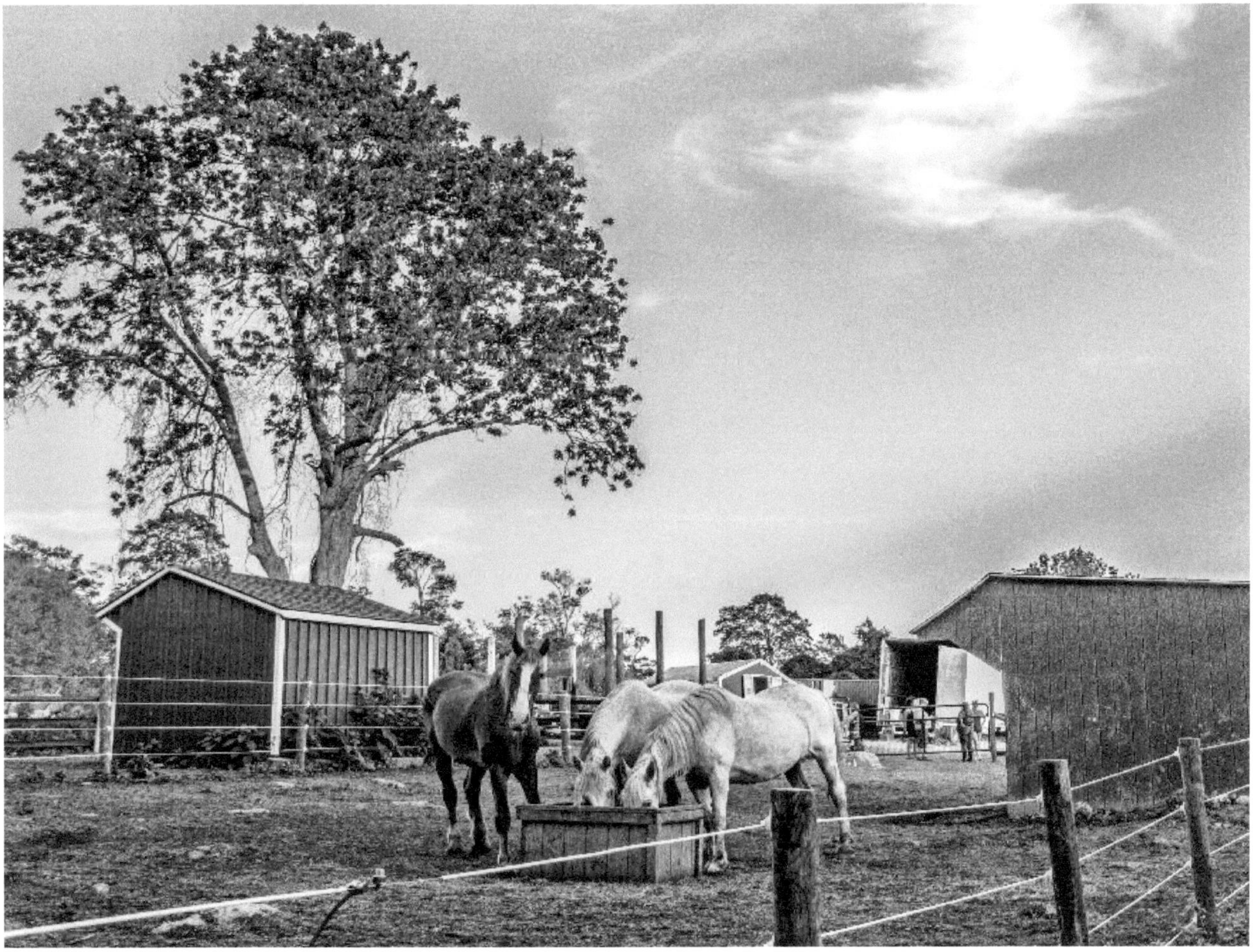

Over time, they made new friends that were not horses or humans. Snakes slithered by them, turtles laid eggs in the paddocks, and birds feasted on the leftover grain that spilled out of the feed pens.

As the months went by, they watched the humans build new things on the farm.

"They must really love us," said Oliver, humbly.

"I hope they know we love them, too" whispered Spencer.

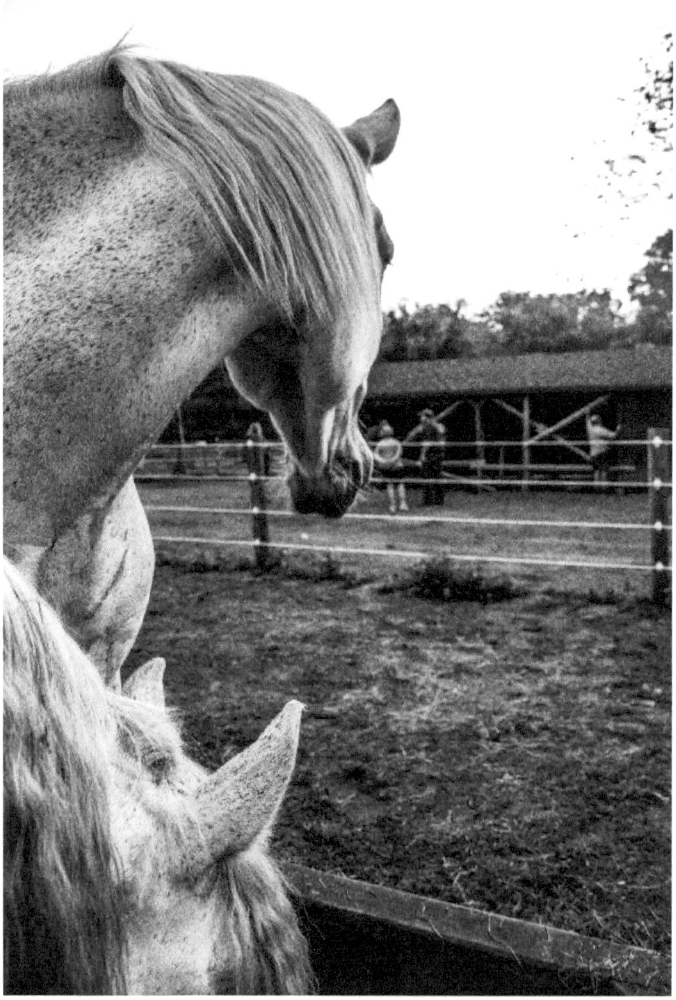

Then one day, their friends brought them to a new training paddock.

"Our ambassadors need to be the first ones to put their hooves in it!" a joyful voice exclaimed.
While in the middle of the paddock, Oliver said, "I'm so proud".

"Me, too" Spencer replied, and then hungrily enjoyed a mouthful of the delicious green grass.

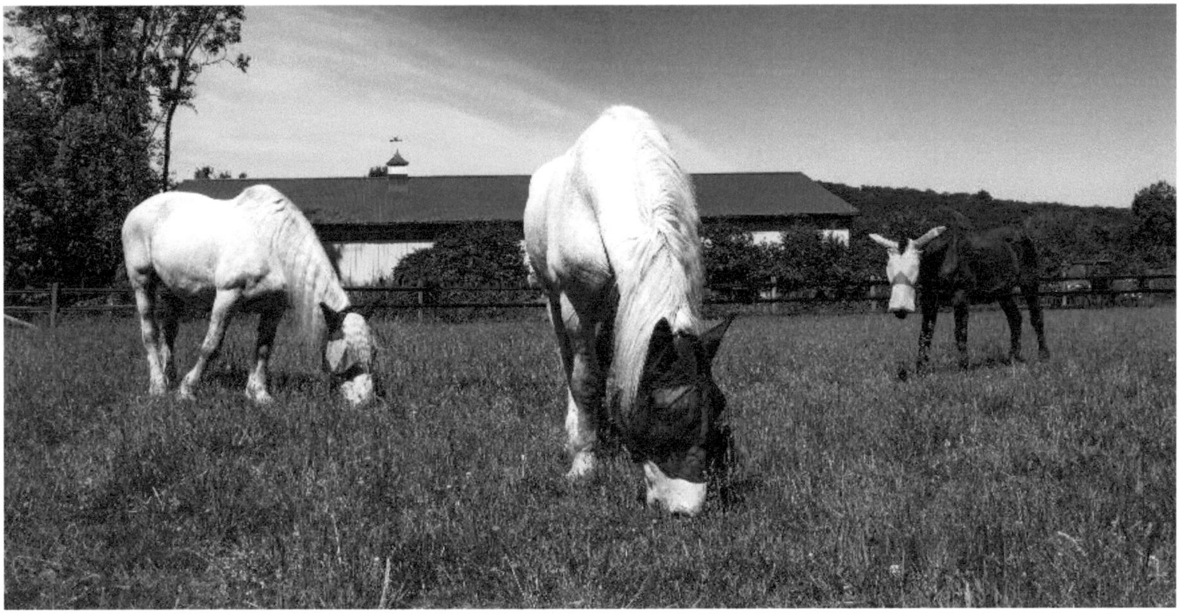

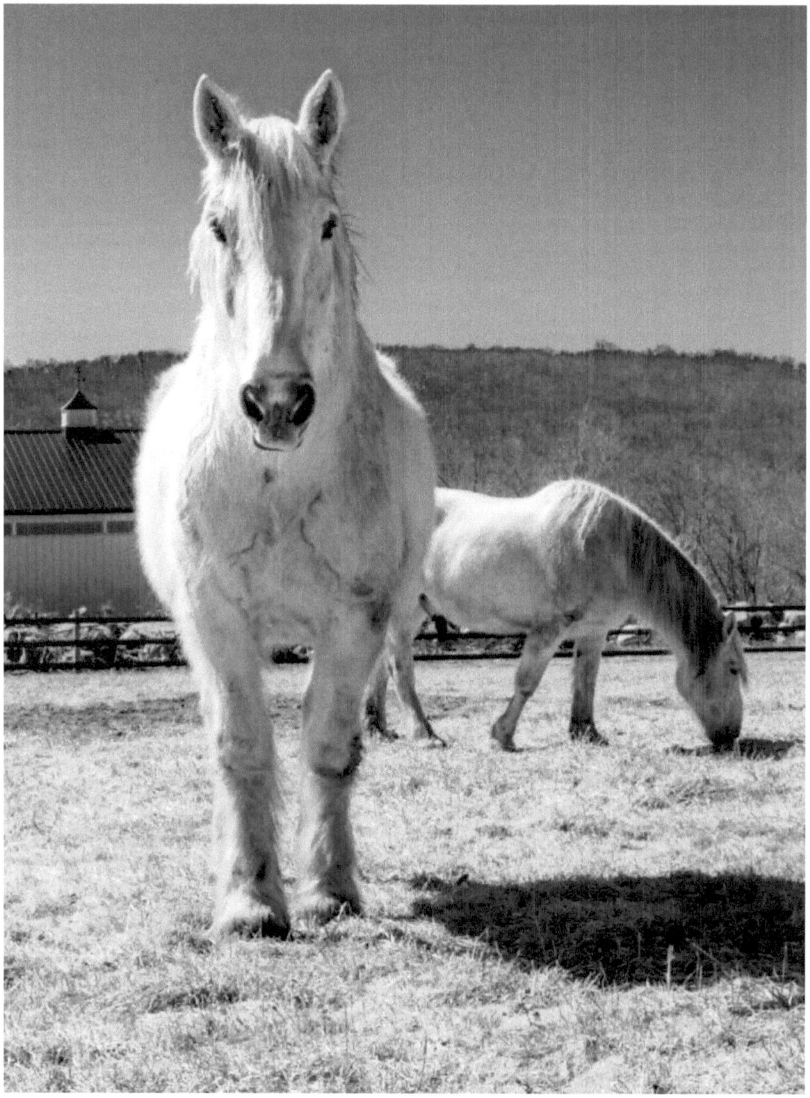

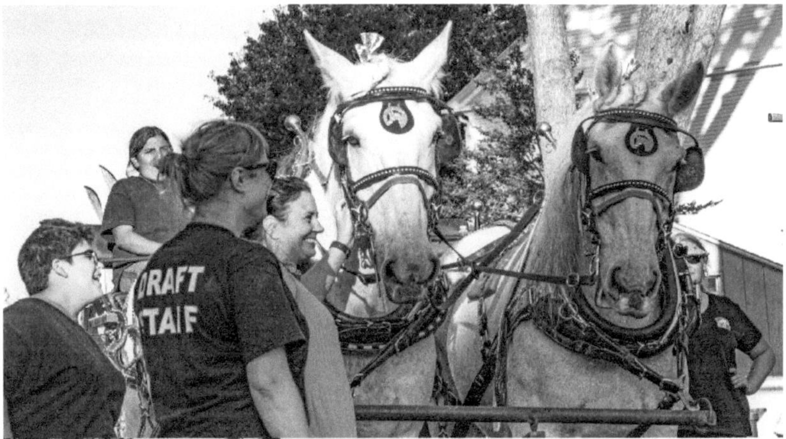

Just when they did not think they could be any prouder, jubilant voices told them "We're going to The Big E!".

Ribbons in manes and tails, clean and shiny tack, the big wagon, and seas of admirers beamed at the Dream Team yet again.

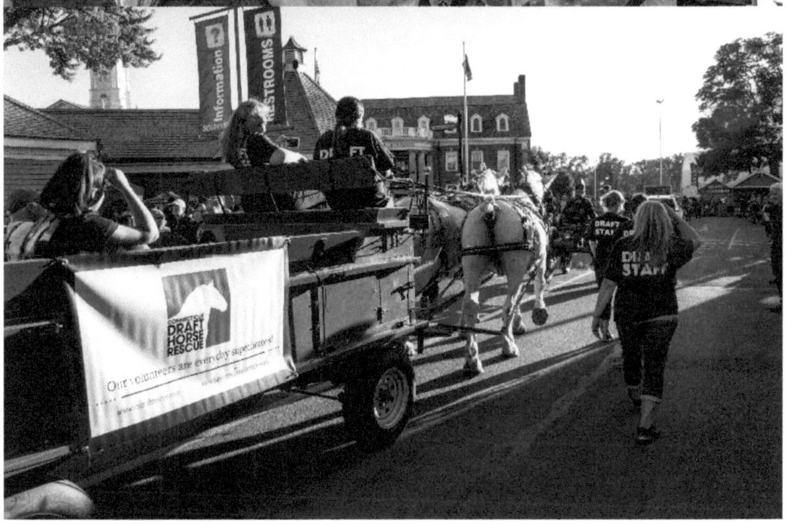

"I sure could use a beer" snickered Spencer at each beer tent.

The trees lost their leaves, many of their bird friends flew south, and the joyous holiday decorations were brought out by the volunteers again.

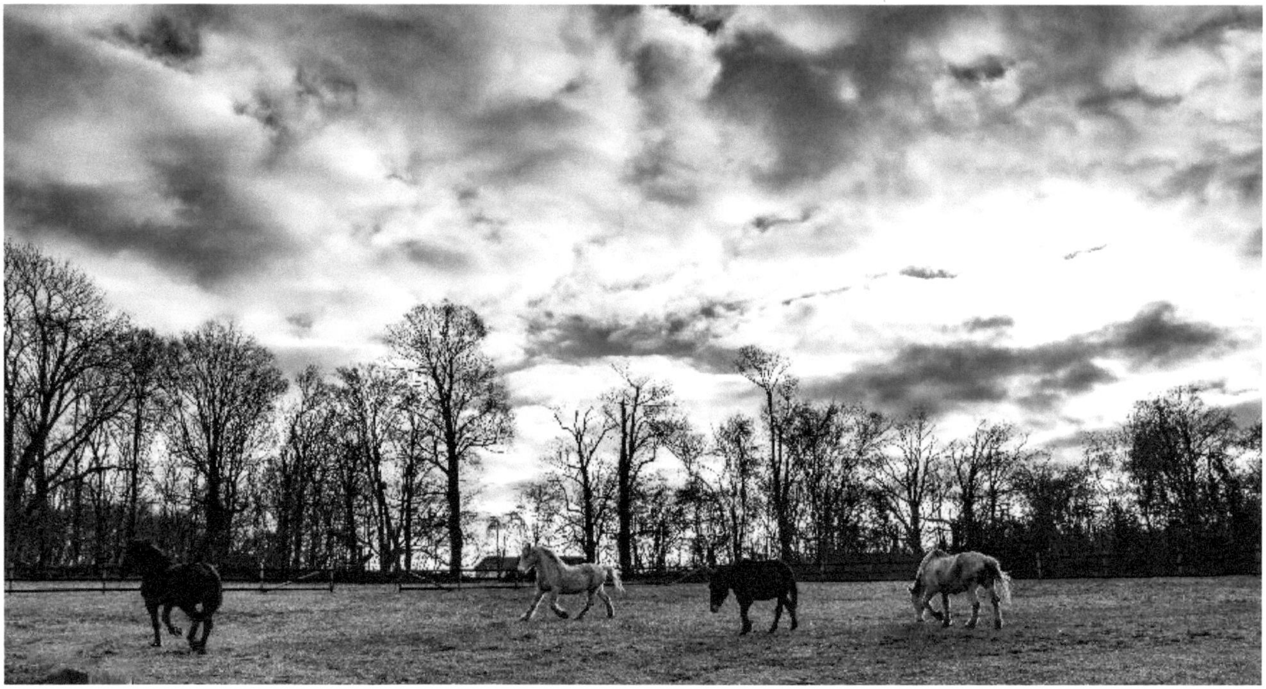

"I feel grumpy this time," said Spencer.

"Well, I'm feeling the Christmas spirit and I'm not afraid of Santa anymore, so I will help you with the Christmas pictures," said Oliver.

"Thanks, friend, you always got my back" replied Spencer.

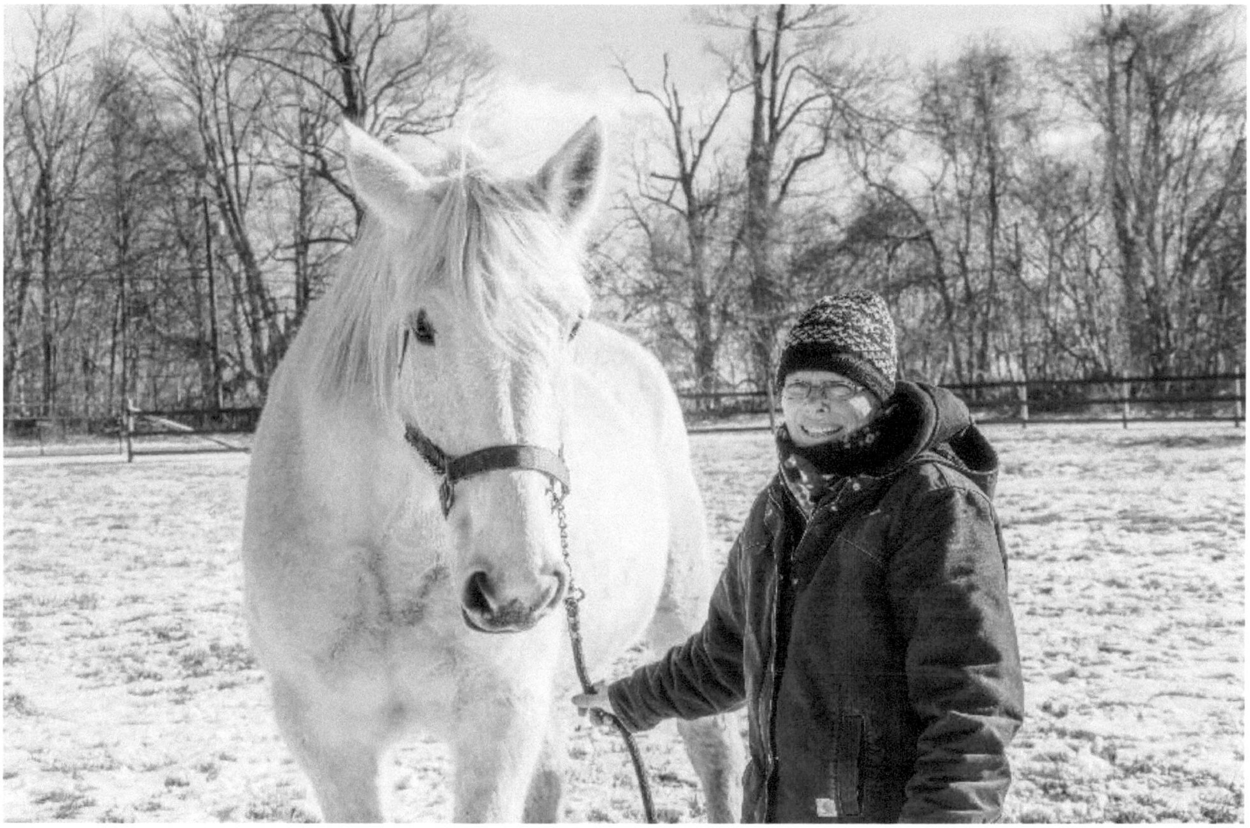

Then, on the winter solstice, a sudden change.
"I cannot get up. Call our human friends!" cried Oliver

Friends encircled, tried to lift, called more friends, and sent an SOS to the group to help the other horses.

"I can't move my legs, but I will keep trying because I love my friends," said Oliver.

"Keep trying" Spencer cried.

"I'm so worried," said Bob.

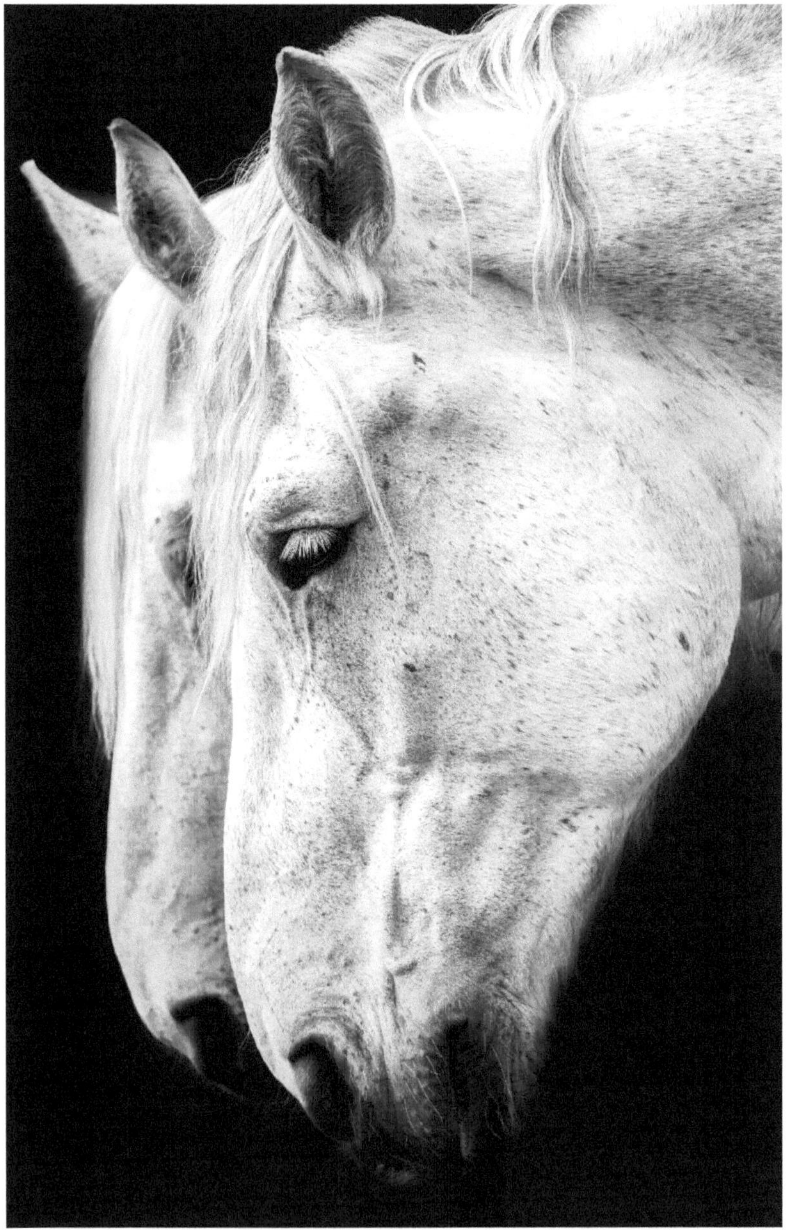

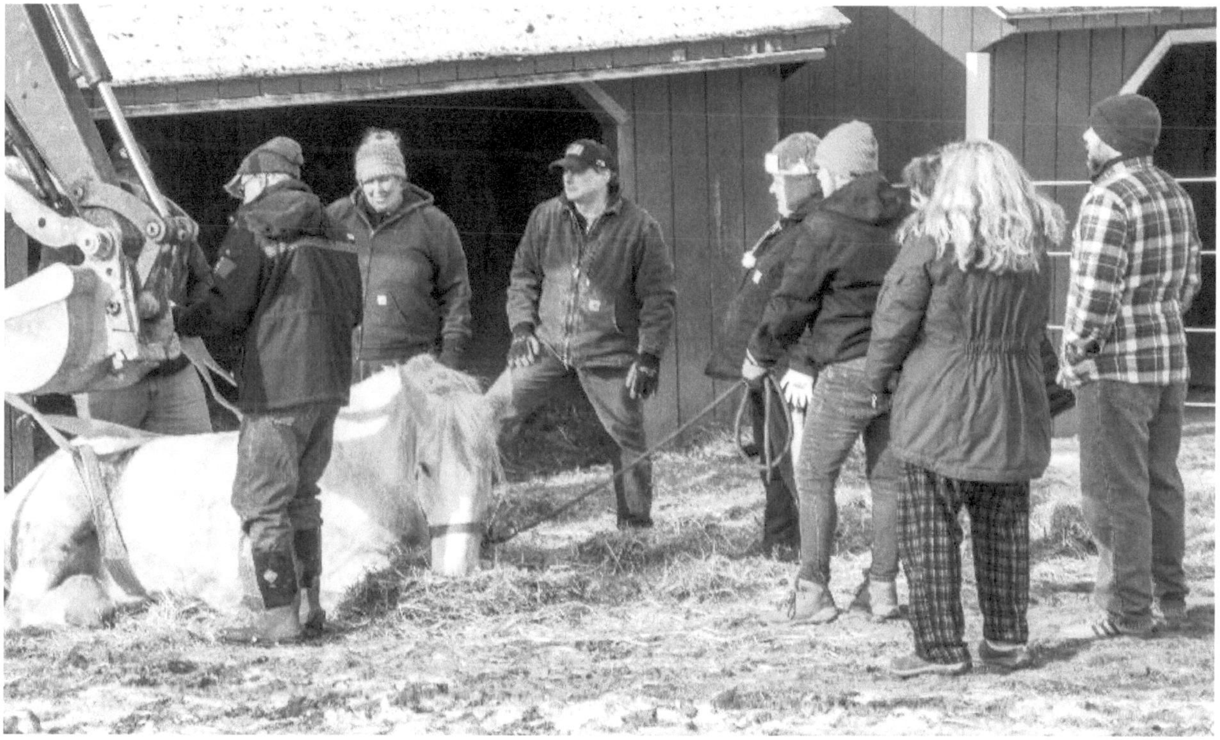

"I feel the love all around me" smiled Oliver, as the volunteers covered him with kisses, hugs, and tears. Even the old man that gave them their first mouthfuls of delicious hay was there to give him one last touch.

All the strength, all the love, all the tears, all the gentle caresses, and all the words of encouragement, could not stop this tragedy.

It was.

Time.

To.

Go.

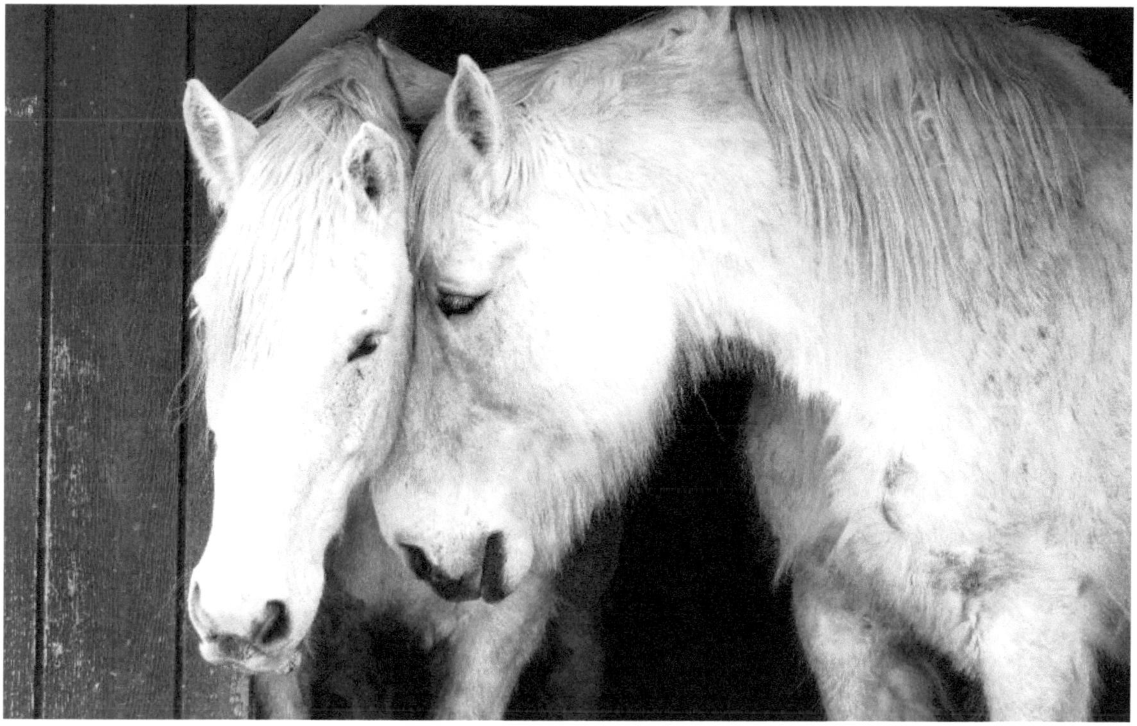

"Goodbye best friend" Spencer wept.

"But you know someone special is about to see you again".

And that is when he knew he would see his original old man at the rainbow bridge.

"I will wait there for you with him," said Oliver.

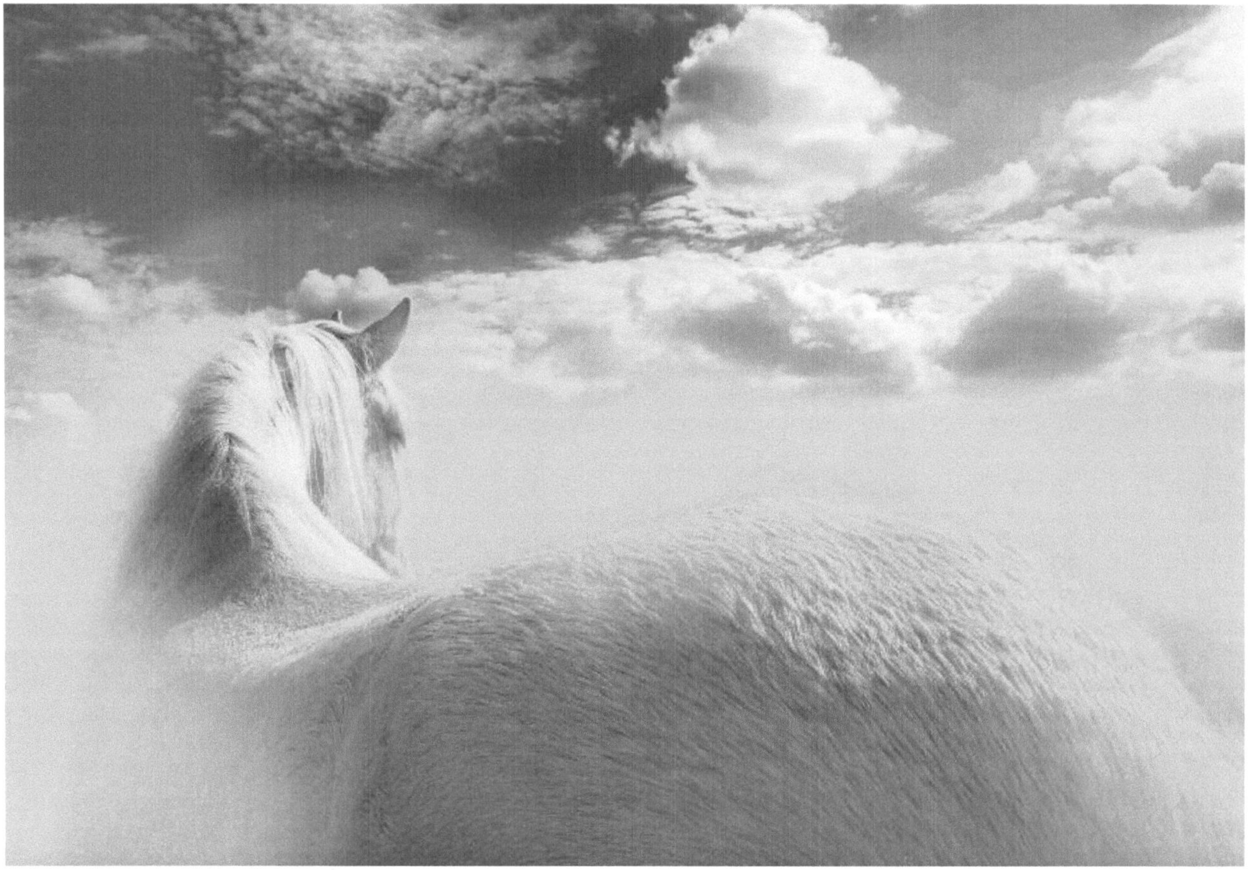

"Wait, I can move again," said Oliver, but he then realized it was because he had gained his wings.

He is the bright blue light you see in the sky. He is the gentle breeze. He is the particles of ash in the training paddock. He is the pride in our hearts.

"I'm still here" mumbled Bob.

"Will you be my new best friend?" asked Spencer.

"You do not mind that I am an old man?" replied Bob.

"Old men are always my best friends. Let me tell you some stories..." Spencer said with a tear and a smile.

They walked to the bottom of the paddock, while the sun was setting above the beautiful Autumn Ridge skyline, while Bob listened to Spencer's glorious stories of the Dream Team.

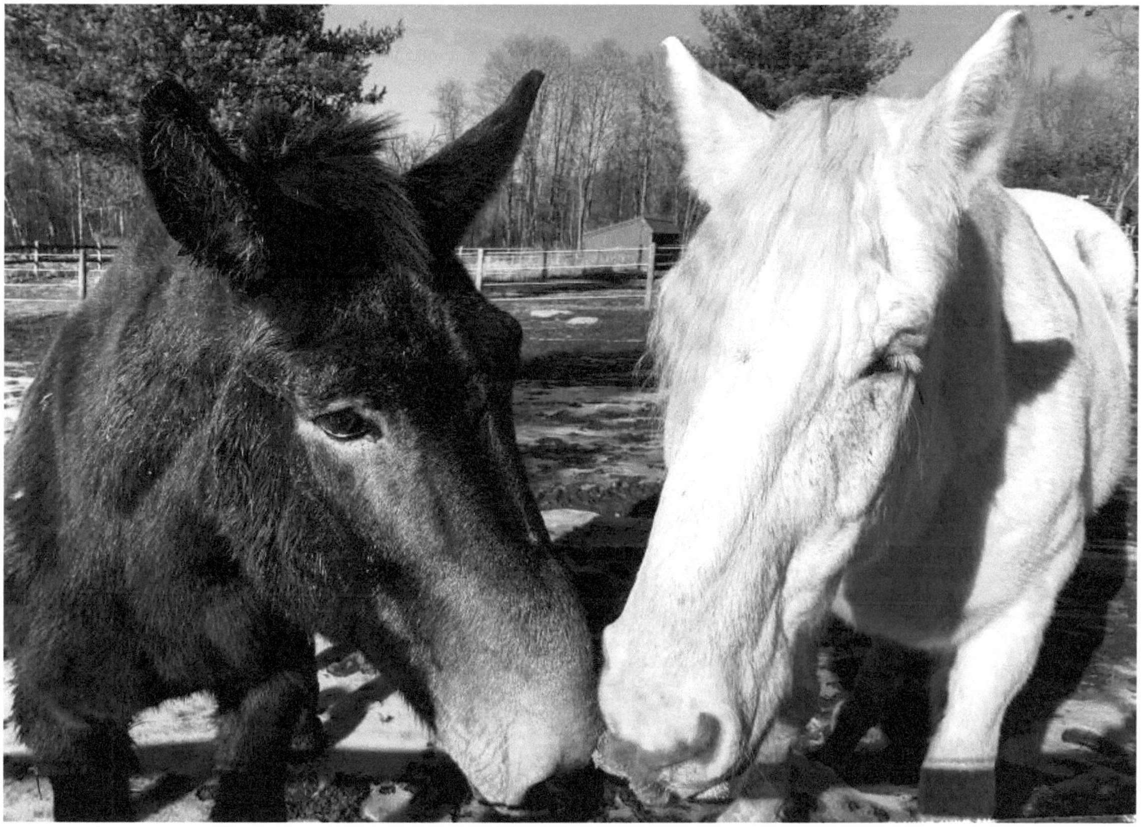

www.ingramcontent.com/pod-product-compliance
Ingram Content Group UK Ltd.
Pitfield, Milton Keynes, MK11 3LW, UK
UKHW061033310726
14089UKWH00010B/226
* 9 7 9 8 9 8 5 3 8 9 1 3 5 *